LOCATION, LOCATION, LOCATION. Much of our lives is spelled out in spatial relationships and physical placement. Specifics of site determine the menu and relative abundance of food for birds, the mating possibilities, the materials available for building nests. The great distances of the frontier West gave rise to another kind of nest, the early trading posts, and encouraged them to become self-contained worlds, cabin-sized embassies of commerce on the foreign soil of native reservations.

The recent work of David Wilcox and his colleagues also turns on physical placement. Looking at the locations of 14th-century ruins across an Arizona mesa, the three archaeologists are reading a tale of war and peace that brings a surprising perspective to our beliefs about early inhabitants of the area.

Steven Carothers describes the impact of human homemaking practices on birds of the Colorado Plateau. He places the Endangered Species Act in a new context for us, as a moral "housekeeping" practice, a secular late-20th-century equivalent to the ancient rituals enacted by native peoples to maintain harmony throughout the nested spheres of the world.

And finally, in her luminous essay "Dwellings," Linda Hogan reminds us that in spite of the physicality of our dwellings, they are also invisible. We reside within them, and they within us. It is of great value to protect our interior dwelling-places, so that we *can* go home again to the hearth within. Perhaps, if enough of our choices are guided by such interior visits, we may come to live in a way that protects the homelands of all creatures, making us better neighbors to those who fly, swim, or run in the delicate dry highlands of the Colorado Plateau.

Dwellings

CAROL HARALSON
AND GREER PRICE

Previous page: Pueblo Bonito. Photo by Clay Martin. Above, left to right: Tent at Toroweap. Photo by Dennis Turville; Mrs. Agular at 121 Benton Street, Flagstaff, Arizona, 1949. Cline Library, Northern Arizona University, NAU.PH.85.3.7.4; Keyhole entrance to prehistoric Puebloan dwelling, southeastern Utah. Photo by Fred Hirschmann. Moonlight Trading Post. Museum of Northern Arizona archives. Bird nest photos by Paul Berkowitz, with special thanks to Elaine Leslie.

ontents

SUMMER 1999

JOINT PUBLISHERS OF PLATEAU JOURNAL

Museum of Northern Arizona

Grand Canyon Association

PLATEAU PARTNERS

Arches National Park

Arizona Strip Interpretive Association

Bryce Canyon Natural History Association

Canyonlands Natural History Association

Capitol Reef National Park

Capitol Reef Natural History Association

Colorado National Monument Association

Dinosaur Nature Association

Dixie Interpretive Association

Entrada Institute/Friends of Capitol Reef

Glen Canyon Natural History Association

Grand Canyon National Park

Grand Canyon Trust

Hubbell Trading Post National Historic Site

Kaibab National Forest

Mesa Verde Museum Association

Mesa Verde National Park

Northern Arizona University, Cline Library

Peaks, Plateaus and Canyons Association

Petrified Forest Museum Association

Petrified Forest National Park

USDI Bureau of Land Management

Zion Natural History Association

PLATEAU JOURNAL
PUBLISHED BY GRAND CANYON
ASSOCIATION AND THE MUSEUM OF
NORTHERN ARIZONA

VOL. 3, NO. 1, SUMMER 1999

THE TRADING POST:
ROADHOUSE OF CULTURE

WITH A PORTFOLIO

OF PALLADIUM PRINTS

BY PAULA JANSEN

Paula Jansen paints with silver and light on paper. Her photographic prints, each a singular work, have timeless faces and surprising depth. We recall the multi-dimensional aspects of life in the tiny outposts called trading posts, strung like space stations across the vast West, with a portfolio of palladium images created by Paula Jansen for *Plateau Journal.*

34

PLATEAU JOURNAL

Editorial and Design
Carol Haralson, Editor and Designer
L. Greer Price, Editor

Pamela Frazier, Partnership Coordinator
Michele Madril, Museum of Northern Arizona
Tracey Hobson, Mesa Verde Museum Association
Paula Branstner, Petrified Forest National Park

Photo Archives and Library
Diane Grua, Cline Library, Northern Arizona University
Tony Marinella, Museum of Northern Arizona
Sara Stebbins, Grand Canyon National Park

Specialized Assistance
Faith Marcovecchio, Grand Canyon Association
Kim Buchheit, Grand Canyon Association
Melissa Canepa Murphy, Marketing and Development

Note: The journal excerpt on page 19 of *Plateau Journal* Vol. 2, No. 2 and accompanying photograph were inadvertently miscredited. Both the journal excerpt and the photograph are by Peg Swift. Our sincere apologies to Ms. Swift.

This publication is made possible in part by support from
the **National Park Foundation**

National Park
FOUNDATION

PRODUCED BY THE MUSEUM OF NORTHERN ARIZONA AND GRAND CANYON ASSOCIATION, *Plateau Journal* is a semiannual publication. Subscription to *Plateau Journal* is a benefit of membership at designated levels in the Museum of Northern Arizona, and in some Plateau Partner organizations.

SUBSCRIPTIONS AND GIFT SUBSCRIPTIONS are available independent of membership. Contact the Museum of Northern Arizona, 520-774-5211, extension 222.

TO PURCHASE COPIES please contact Grand Canyon Association, 800-858-2808. Single copies are available for $9.95 and special discounts are available for quantity purchases. Back issues are available.

PLEASE ADDRESS CORRESPONDENCE TO *Plateau Journal,* Museum of Northern Arizona, 3101 N. Fort Valley Road, Flagstaff, AZ 86001. Telephone 520-774-5211, extension 216.

Printed on recycled paper with soy-based inks.
Printed in Canada by Friesen Printing.

Bird nest. Photo by Paul Berkowitz, with special thanks to Elaine Leslie. Facing: Aerial view of Perry Mesa. Photo by Jerry Jacka; Frances Ruin Complex (Navajo). Photo by Clay Martin.

44

PERRY MESA, A 14TH-CENTURY GATED COMMUNITY IN CENTRAL ARIZONA

DAVID R. WILCOX, GERALD ROBERTSON, JR.

AND J. SCOTT WOOD

AERIAL PHOTOGRAPHY BY JERRY JACKA

In the mid-14th century, inhabitants of central Arizona built a constellation of settlements whose locations tell a fascinating story. The relationships between land and settlement, and between one settlement and another, are a tangible expression of fear and aggression, of cooperation and unity. Working with Vietnam War-trained avocational archaeologist Gerald Robertson, David Wilcox of the Museum of Northern Arizona and Scott Wood of Tonto National Forest are interpreting the spatial language of this tale spelled in stone.

"*Are we perhaps on the verge of grasping that
the environment is ourselves, for it has given
us form, and that creation is nothing but a
dialogue between the inside and the outside?*"

— GASTON BACHELARD, *The Poetics of Space*

DWELLINGS

And so when we examine a nest, we place ourselves at the origin of confidence in the world, we receive a beginning of confidence, an urge toward cosmic confidence. Would a bird build its nest if it did not have its instinct for confidence in the world?

GASTON BACHELARD,
THE POETICS OF SPACE

NOT FAR FROM WHERE I LIVE is a hill that was cut into by the moving water of a creek. Eroded this way, all that's left of it is a broken wall of earth that contains old roots and pebbles woven together and exposed. Seen from a distance, it is only a rise of raw earth. But up close it is something wonderful, a small cliff dwelling that looks almost as intricate and well made as those the Anasazi left behind when they vanished mysteriously centuries ago. This hill is a place that could be the starry skies of night turned inward into the thousand round holes where solitary bees have lived and died. It is a hill of tunneling rooms. At the mouths of some of the excavations, half-circles of clay beetle out like awnings shading a doorway. It is earth that was turned to clay in the mouths of the bees and spit out as they mined deeper into their dwelling places.

This place where the bees reside is at an angle safe from rain. It faces the southern sun. It is a warm and intelligent architecture of memory, learned by whatever memory lives in the blood. Many of the holes still contain the gold husks of dead bees, their faces dry and gone, their flat eyes gazing out from death's land toward the other uninhabited half of the hill that is across the creek from these catacombs.

The first time I found the residence of the bees, it was dusty summer. The sun was hot, and

Above: Cliff swallow nests above pictographs near Pueblo Penasco, Chaco Culture National Historic Park, New Mexico. Right: Keet Seel, prehistoric Pueblo cliff dwelling, Navajo National Monument, Arizona. Photos by Fred Hirschmann.

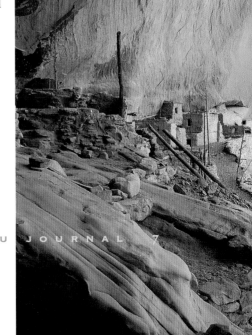

land was the dry color of rust. Now and then a car rumbled along the dirt road and dust rose up behind it before settling back down on older dust. In the silence, the bees made a soft droning hum. They were alive then, and working the hill, going out and returning with pollen, in and out through the holes, back and forth between daylight and the cooler, darker regions of inner earth. They were flying an invisible map through air, a map charted by landmarks, the slant of light, and a circling story they told one another about the direction of food held inside the center of yellow flowers.

S ITTING IN THE HOT SUN, watching the small bees fly in and out around the hill, hearing the summer birds, the light breeze, I felt right in the world. I belonged there. I thought of my own dwelling places, those real and those imagined. Once I lived in a town called Manitou, which means "Great Spirit," and where hot mineral springwater gurgled beneath the streets and rose up into open wells. I felt safe there. With the underground movement of water and heat a constant reminder of other life, of what lives beneath us, it seemed to be the center of the world.

A few years after that, I wanted silence. My daydreams were full of places I longed to be, shelters and soltiudes. I wanted a room apart from others, a hidden cabin to rest in. I wanted to be in a redwood forest with trees so tall the owls called out in the daytime. I daydreamed of living in a vapor cave a few hours away from here. Underground, warm, and moist, I thought it would be the perfect world for staying out of cold winter, for escaping the noise of living.

And how often I've wanted to escape to a wilderness where a human hand has not been in everything. But those were only dreams of peace, of comfort, of a nest inside stone or woods, a sanctuary where a dream or life wouldn't be invaded.

Y EARS AGO, IN THE NEXT CANYON WEST OF HERE, there was a man who followed one of those dreams and moved into a cave that could only be reached by climbing down a rope. For years he lived there in comfort, like a troglodite. The inner weather was stable, never too hot, too cold, too wet, or too dry. But then he felt lonely. His utopia needed a woman. He went to town until he found a wife. For a while after the marriage, his wife climbed down the rope along with him, but before long she didn't want the mice scurrying about in the cavern, or the untidy bats that wanted to hang from stones of the ceiling. So they built a door. Because of the closed entryway, the temperature changed. They had to put in heat. Then the inner moisture of earth warped the door, so they had to have air-conditioning, and after that the earth wanted to go about life in its own way and it didn't give in to the people.

Above: Woodrat in the entrance to its home.
©Marty Cordano/DRK PHOTO.

Sunlight streams toward the painted inner wall of a prehistoric dwelling, southeastern Utah. Photo by Fred Hirschmann.

". . . the houses that were lost forever . . .

continue to live on in us"

— GASTON BACHELARD, *The Poetics of Space*

IN OTHER DAYS AND PLACES, people paid more attention to the strong-headed will of earth. Once homes were built of wood that had been felled from a single region in a forest. That way, it was thought, the house would hold together more harmoniously, and the family of walls would not fall or lend themselves to the unhappiness or arguments of the inhabitants.

AN ITALIAN IMMIGRANT TO CHICAGO, Aldo Piacenzi, built birdhouses that were dwellings of harmony and peace. They were the incredible spired shapes of cathedrals in Italy. They housed not only the birds, but also his memories, his own past. He painted them the watery blue of his Mediterranean, the wild rose of flowers in a summer field. Inside them was straw and droppings of lives that layed eggs, fledglings who grew there. What places to inhabit, the bright and sunny birdhouses in dreary alleyways of the city.

ONE BEAUTIFUL AFTERNOON, cool and moist, with the kind of yellow light that falls on earth in these arid regions, I waited for barn swallows to return from their daily work of food gathering. Inside the tunnel where they live, hundreds of swallows had mixed their saliva with mud and clay, much like the solitary bees, and formed nests that were perfect as a potter's bowl. At five in the evening, they returned all at once, a dark, flying shadow. Despite their enormous numbers and the crowding together of nests, they didn't pause for even a moment before entering the nests, nor did they crowd one another. Instantly, they vanished into the nests. The tunnel went silent. It held no outward signs of life.

But I knew they were there, filled with the fire of living. And what a marriage of elements was in those nests. Not only mud's earth and water, the fire of sun and dry air, but even the elements contained one another. The bodies of prophets and crazy men were broken down in that soil.

I'VE NOTICED OFTEN how when a house is abandoned, it begins to sag. Without a tenant, it has no need to go on. If it were a person, we'd say it is depressed or lonely. The roof settles in, the paint cracks, the walls and floorboards warp and slope downward in their own natural ways, telling us that life must stay in everything as the world whirls and tilts and moves through boundless space.

ONE SUMMER DAY, cleaning up after long-eared owls where I work at a rehabilitation facility for birds of prey, I was raking the gravel floor of a flight cage. Down on the ground, something looked like it was moving. I bent over to look into the pile of bones and pellets I'd just raked together. There, close to the ground, were two fetal mice. They were new to the planet, pink and hairless. They were so tenderly young. Their faces had swollen blue-veined eyes. They were nestled in a mound of feathers, soft as velvet, each one curled up smaller than an infant's ear, listening to the first sounds of earth. But the ants were biting them. They turned in agony, unable to pull away, not yet having the arms

"The enterprise and skill with which animals make their nests is so efficient that it is not possible to do better, so entirely do they surpass all masons, carpenters and builders; for there is not a man who would be able to make a house better suited to himself and his children. . ."

— AMBROSE PARÉ

Thomas family in front of their home, Oak Creek, Arizona. May Hicks Curtis Hill Collection, Cline Library, Northern Arizona University, NAU.PH.91.7.387.

Facing: Left, a Navajo woman prepares a meal. NAU.PH.93.6.4. Right, Tewa girls, Hopi Indian Reservation, Arizona. Edward S. Curtis Collection, NAU.PH.93.38.31. Both courtesy Cline Library, Northern Arizona University.

or legs to move, but feeling, twisting away from, the pain of the bites. I was horrified to see them bitten out of life that way. I dipped them in water, as if to take away the sting, and let the ants fall in the bucket. Then I held the tiny mice in the palm of my hand. Some of the ants were drowning in the water. I was trading one life for another, exchanging the lives of ants for those of mice, but I hated their suffering, and hated even more that they had not yet grown to a life, and already they inhabited the miserable world of pain. Death and life feed each other. I know that.

Inside these rooms where birds are healed, there are other lives besides those of mice. There are fine gray globes the wasps have woven together, the white cocoons of spiders in a corner, the downward tunneling anthills. All these dwellings are inside one small walled space, but I think most about the mice. Sometimes the downy nests fall out of the walls where their mothers have placed them out of the way of their enemies. When one of the nests falls, they are so well made and soft, woven mostly from the chest feathers of birds. Sometimes the leg of a small quail holds the nest together like a slender cornerstone with dry, bent claws. The mice have adapted to life in the presence of their enemies, adpated to living in the thin wall between beak and beak, claw and claw. They move their nests often, as if a new rafter or wall will protect them from the inevitable fate of all our returns home to the deeper, wider nest of earth that houses us all.

Racoons in their den.
©C.C. Lockwood/DRK PHOTO.

ONE AUGUST AT ZIA PUEBLO during the corn dance I noticed tourists picking up shards of all the old pottery that had been made and broken there. The residents of Zia know not to take the bowls and pots left behind by the older ones. They know that the fragments of those earlier lives need to be smoothed back to earth, but younger nations, travelers from continents across the world who have come to inhabit this land, have little of their own to grow on. The pieces of earth that were formed into bowls, even on their way home to dust, provide the new people a lifeline to an unknown land, help them remember that they live in the old nest of earth.

IT WAS IN EARLY FEBRUARY, during the mating season of the great horned owls. It was dusk, and I hiked up the back of a mountain to where I'd heard the owls a year before. I wanted to hear them again, the voices so tender, so deep, like a memory of comfort. I was halfway up the trail when I found a soft, round nest. It had fallen from one of the bare-branched trees. It was a delicate nest, woven together of feathers, sage, and strands of wild grass. Holding it in my hand

". . . the house shelters

daydreaming, the house protects

the dreamer, the house allows one

to dream in peace."

— GASTON BACHELARD,

The Poetics of Space

Window, Jail House Ruin, Bullet Canyon, Grand Gulch Area, Utah. Photo by Fred Hirschmann.

in the rosy twilight, I noticed that a blue thread was entwined with the other gatherings there. I pulled at the thread a little, and then I recognized it. It was a thread from one of my skirts. It was blue cotton. It was the unmistakable color and shape of a pattern I knew. I liked it, that a thread of my life was in an abandoned nest, one that had held eggs and new life. I took the nest home. At home, I held it to the light and looked more closely. There, to my surprise, nestled into the gray-green sage, was a gnarl of black hair. It was also unmistakable. It was my daughter's hair, cleaned from a brush and picked up out in the sun beneath the maple tree, or the pit cherry where birds eat from the overladen, fertile branches until only the seeds remain on the trees.

Private residence in Utah, built into cavern wall. Photo by Dennis Turville.

I didn't know what kind of nest it was, or who had lived there. It didn't matter. I thought of the remnants of our lives carried up the hill that way and turned into shelter. That night, resting inside the walls of our home, the world outside weighed so heavily against the thin wood of the house. The sloped roof was the only thing between us and the universe. Everything outside of our wooden boundaries seemed so large. Filled with night's citizens, it all came alive. The world opened in the thickets of the dark. The wild grapes would soon ripen on the vines. The burrowing ones were emerging. Horned owls sat in treetops. Mice scurried here and there. Skunks, fox, the slow and holy porcupine, all were passing by this way. The young of the solitary bees were feeding on pollen in the dark. The whole world was a nest on its humble tilt, in the maze of the universe, holding us.

Bedstead at Bill Bass's trailside camp on Bass Trail, Grand Canyon. Photo by Paul Berkowitz.

LINDA HOGAN is a well-known author of both fiction and nonfiction. Her latest novel, *Power,* was published in 1998. She is also the editor of *Intimate Nature: The Bond Between Women and Animals,* an anthology of nature writing featuring the collected works of more than seventy women. Her essay "Dwellings" appeared in a collection of her writings entitled *Dwellings: A Spiritual History of the Living World,* published by Norton in 1995. She lives in Colorado.

Edith Bass feeding her chickens at the Grand Canyon, ca. 1910. Cline Library, Northern Arizona University, NAU.PH.96.3.3.5.

Ladder into cliffside cave, Bandelier National Monument, New Mexico. Photo by S. Brooks Bedwell.

Spirit Guides, Prayer Feathers, and Omens of Darkness

Endangered Birds of the Colorado Plateau

In his classic *Sun Chief: Autobiography of a Hopi Indian,* Don Talayesva tells us how he learned from his grandfather that

MY GRANDFATHER TOLD ME THAT ALL THE SONGBIRDS ARE THE SUN GOD'S PETS, PUT HERE TO KEEP THE PEOPLE HAPPY WHILE THEY WORK.

SUN CHIEF, 1942

birds are messengers from the gods. I know this to be true. It was the spotted owl who told us that management for even-aged stands of pine trees was choking life from the land. I believe that owl has another message, telling us to ask the sheep and cows to leave at least a little of the understory vegetation for brother rat, so that owl may eat and live. The bald eagle and peregrine falcon also delivered their warnings; fortunately scientists interpreted the message, and the use of deadly chlorinated hydrocarbon pesticides was discontinued. These same raptors tell us, too, that they are able to live in a technologically modified world that is ever changing, that they can live sympatrically within human-modified habitats, if those habitats meet their needs. I believe that the influence of birds on the human psyche has contributed powerfully to the effort to keep human-caused extinctions of *all* species in check, mostly through regulatory and moral support for the Endangered Species Act.

Facing: Aspen in autumn light. Photo by S. Brooks Bedwell.
Above: Feather *paho* (prayer stick). Photo by Liz Hymans.

Water droplets on lupine, Wasatch National Forest, Utah. Photo by Jerry Sintz.

It is deep winter now on the southern flanks of the Colorado Plateau near Flagstaff, and the birds are mostly gone. Except for the occasional piercing cries of foraging Stellar's jays and Clark's nutcrackers, the ponderosa forest is silent. The only sign I see of spring and summer birds are feathers of *sikya'tsi,* the yellow-warbler, attached to a *paho,* or prayer feather, on my desk. Like all prayers, the *paho,* a gift from a Hopi friend, is an offering to the Great Spirit, in this case on my behalf. The offering is a precious thought, symbolized by the feathers and their flight towards the spirits, that benefits both the giver and receiver.

The avifauna of the Colorado Plateau have a special place in the myth and story of all Native American peoples of the Colorado Plateau. After thirty-five years of studying and living within the wilds of the Colorado Plateau, I find that the birds, especially certain species, have come to be important symbols of my unconscious communications with those same spirits. It is not just the Native Americans of our environs who can claim multigenerational connections to the land. My grandson Jacob's great-great-great grandfather Dr. Warren E. Day, a horse-and-buggy healer, worked among the Yavapai, Apache, and Hualapai from his home, my birthplace, in Prescott, Arizona. I have a strong sense of belonging to this country, of always having been here.

Traveling throughout the plateau has provided the opportunity to feel firsthand the wonders of wilderness and to witness what I have always seen as ecological balance. Isolated regions of the Colorado Plateau remain today much as they were hundreds, if not thousands, of years ago. Yet even here evidence of ecosystem decay provides ominous harbingers of unintended and unexpected consequences of the human animal's population growth, technology, and settlement patterns. Throughout the years I have studied plateau wildlife, the birds, especially those species we now recognize as threatened or endangered, have been my "spirit guides." They have shaped my ideas about where we as humans are going and who we will "take with us" or allow to live in peace in a world characterized by my Native American friends' concept of harmony among all things.

Cloud reflections in rainwater pool, Canyonlands National Park, Utah. Photo by S. Brooks Bedwell.

My travels and studies have brought friendships with Navajos, Hopis, Apaches, Hualapais, and Paiutes who have given me a non-scientific way of looking at the workings of the world. A central theme pervasive within their philosophies is that life, human life especially, turns on the quest for harmony and balance among all natural things. If the rituals and ceremonies enacted over generations are not followed precisely, if humans act in selfish ways counter to the teachings of their elders, disharmony will result. In individuals, this may translate into sickness, injury, or misfortune. For Mother Earth, expressions of disharmony are drought, catastrophic forest fires, floods, pestilence, and similar visitations of disaster. From my limited understanding of other cultures both ancient and modern, on the plateau and beyond, I suspect a similar world view underpins our broader, not-so-traditional society: violate ancient teachings, behave selfishly, and life force systems will decay into disharmony. Act thoughtfully, with a view toward the larger and long-term implications of our actions, and balance can be restored.

I see direct parallels here with our changing perspective on the natural world. The science of ecology has evolved to allow much greater understanding of the cause-and-effect relationship between human actions and disruptions in the dynamic ecological balance of nature. One outcome of such disruption is the decline and loss of species. Traditional Native American methods for restoring balance involved ritual, fasting, and prayer; modern society works to restore balance in biodiversity through legislation, particularly the Endangered Species Act.

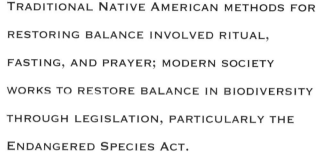

TRADITIONAL NATIVE AMERICAN METHODS FOR RESTORING BALANCE INVOLVED RITUAL, FASTING, AND PRAYER; MODERN SOCIETY WORKS TO RESTORE BALANCE IN BIODIVERSITY THROUGH LEGISLATION, PARTICULARLY THE ENDANGERED SPECIES ACT.

On the Colorado Plateau, we recognize that a number of plant and animal species are experiencing serious population declines. There is no question that species were appearing and disappearing long before the human animal leapt from the trees to the moon. In the Grand Plan of Nature, species come and species go. When I speak of serious population declines of Colorado Plateau wildlife, I am referring only to those linked directly to you and me, *Homo sapiens*. Although there may be dozens of birds on the Colorado Plateau (out of the over 400 species known to use the area) for which reliable data indicate declining numbers over the past several decades, only nine species are officially listed by the U.S. Fish and Wildlife Service as in danger of becoming extinct. Principal among these species are the California condor, American peregrine falcon, bald eagle, and Mexican spotted owl.[+] These birds and their plateau dwelling places have played a major role in my professional career and, in a larger sense, provided guidance in the way I comport myself.

From this perspective therefore, with only occasional lapses in my training as an empirical scientist, I will tell the plateau stories of falcon, eagle, condor, and owl, and of human efforts to spare them from annihilation. On the whole, I see these species as better off today than when my studies started some three decades ago. They are better off in some cases because their numbers have actually increased, but also because, in a more general sense, an innate need to live in equilibrium, in harmony with the surrounding world, seems to be reasserting itself in modern humans after a long period of quiescence. Most of us cannot articulate precisely why we don't want the condor to go extinct, why the old-growth forests should be allowed to stand, why the sight of a bald eagle carrying a fish can be a numinous experience. All we know is that these elements of our world should be preserved. If the ecological balance of their habitats is disrupted, balance should be restored.

FEW PEOPLE WERE AWARE of the role of *Homo sapiens* in the loss of species before the last wild passenger pigeon was shot out of North American skies at the beginning of this century. The disappearance of that lamented species, once the most numerous on Earth, propelled the issue of human-caused extinction beyond the ken of a few into the general consciousness. This awareness gradually grew as more and more species declined and vanished before the sweep of industrialized society. Saddened at the loss of our fellow travelers on this planet, and fearful of the consequences to our own species, Americans began to seek a way to stem, even reverse, the trend. Unlike the ritual and myth used by traditional cultures to maintain stability within their natural environment, our modern tools of collective action are statutes, regulations, and governmental bureaucracies. Hence the Endangered Species Act of 1973.

Other laws have been passed at both the federal and state levels to preserve species—notably (for birds) the Migratory Bird Treaty Act of 1918, the Bald Eagle Protection Act of 1940, and the Wild Bird Conservation Act of 1992—but the Endangered Species Act is the cornerstone of this legislative edifice. The point

THERE IS NO QUESTION THAT SPECIES WERE APPEARING AND DISAPPEARING LONG BEFORE THE HUMAN ANIMAL LEAPT FROM THE TREES TO THE MOON. . . . WHEN I SPEAK OF SERIOUS POPULATION DECLINES . . . I AM REFERRING ONLY TO THOSE LINKED DIRECTLY TO YOU AND ME . . .

+ The other federally listed species associated with Colorado Plateau environs are the southwestern willow flycatcher, arctic peregrine falcon, brown pelican, whooping crane, and least tern. Several more species of birds are given special status and some measure of special treatment by various state wildlife agencies and individual federal agencies.

REDRESSING IMBALANCE:

THE ENDANGERED SPECIES ACT

Facing: Cactus wren nesting in cholla. Photo by T.C. Brown.

A young condor at Horseshoe Bend, Glen Canyon, Arizona. Olfactory clues can lead condors to food, but this portajohn, though it smelled right, was not supper. Photo by Mike Boyle.

of the act is to prevent species from going extinct by identifying, or listing, those whose existence is in jeopardy and by protecting habitat deemed essential for their survival and recovery. Identified species may be listed as endangered (in danger of extinction) or threatened (likely to become endangered within the foreseeable future). Among other provisions designed to protect listed creatures and their habitats, the act requires that U.S. Fish and Wildlife Service develop and implement recovery plans for the species—sets of actions likely to promote population growth. When species recover sufficiently, they can be removed from the endangered list, a process called delisting.

Currently, the California condor and American peregrine falcon are listed as endangered, although falcons have increased to such an extent that they have been proposed for delisting. The bald eagle and the Mexican spotted owl are listed as threatened. Each of these four species was or is in danger of extinction. Each has been the subject of focused and costly efforts to ensure its survival. In the cases of the peregrine falcon and bald eagle, the results have been resoundingly positive. For the Mexican spotted owl, the prospect is also relatively sanguine, though it is too early to tell; so far this bird appears to be holding its own. But the magnificent California condor skirts the edge of extinction. Within the past thirty years, I have been involved in sometimes casual, sometimes detailed studies on all of these species. And when not studying the birds, I have always been aware of their presence and their importance to Colorado Plateau ecosystem diversity. Here are my stories of each of these birds as I have encountered them in our common dwelling place, the Colorado Plateau.

CALIFORNIA CONDOR

WHEN GREER CHESHER provided a hopeful article to the Museum of Northern Arizona publication *Cañon Journal* detailing a plan for "returning" the endangered California condor to the skies of the Colorado Plateau, my reaction to the plan was, I have to admit, pessimistic. This was despite the fact that I had advocated this very thing several years before.

My first interaction with condor biologists occurred in California in the early 1980s when a scientific debate raged over the advisability of removing the last free-flying condors from the wild. The remaining dozen-plus adults at that time inhabited an area north of Los Angeles. Supporters of removal argued that the only way to save the species was to protect the last of the available "brood stock" by capturing all the adults and captive-rearing their young for later release. Allowing these birds to remain in the wild was folly, these biologists argued, because condors were systematically dying in a habitat that was becoming more and more inhospitable.

Inadequate food to maintain viable populations of these gigantic scavengers probably initiated their decline, but in the early 1980s the immediate dangers were poisoning from ingested bullets, shotgun pellets, pesticides, and other carrion contaminants as well as shootings and other human-related hazards. At that time, lead poisoning was pervasive throughout the remaining California condors because the only available food left for them included deer that had been wounded during the hunting season. Condors, finding and feeding on the carcasses, ingested bullets and bullet fragments. Lead accumulated in their bodies, poisoning them.

California condor at Grand Canyon.
Photo by Marty Cordano.

Opponents of capturing the remaining birds cited problems associated with propagation in captivity and habituating animals to human presence. Some expressed philosophical, ethical, and aesthetic concerns about imprisoning the last of these great, far-ranging creatures, the largest birds in North America. Let the species survive or die, they said, as free-flying, wild animals.

While the disquisition over the condor's fate was evolving, I was part of a team analyzing the environmental consequences of developing a wind energy field near the town of Gorman, California. The proposal to place this relatively environmentally friendly, albeit ugly, field of 500 windmills on a windy California hillside was ultimately defeated because of the condor. At that time, a few condors still occupied the region, particularly the nearby Sespe Wilderness area, and biologists spoke of possibly reintroducing captive-bred condors there in the future. Many condor biologists and their environmental activist supporters expressed concern that, if the condor were to return, they could be killed or maimed in collisions with the windmills.

When the last public meeting over the windmills was held in Gorman, I suggested that if the condor recovery group wanted to release healthy condors somewhere, they should consider Grand Canyon country. Releasing them into the same habitat that was causing their demise was, in my opinion, absurd. I even went so far as to propose to the National Park Service and Hualapai Tribe (whose lands include portions of western Grand Canyon) that they become involved in the effort. I believed that the Grand Canyon, which falls within the historic California condor range, would provide the winds, cliffs, water, and isolation

Vermilion Cliffs. Photo by Liz Hymans.

needed by these birds. Sufficient food, long an issue for this species, may also have been available. A large number of feral burros occupy Hualapai land, damaging rangeland for both livestock and wildlife. Slaughtering environmentally unfriendly burros to feed endangered condors would kill two birds with one stone, as the metaphor goes. While setting up such an unnatural feeding situation was far from ideal in my view, the Grand Canyon still seemed to be a better choice than an area proven to be deleterious to the species. Hualapai Tribe wildlife biologist Clay Bravo and I, however, were never able to generate interest in our proposal.

California condors made it to the Grand Canyon nonetheless. Once U.S. Fish and Wildlife Service personnel and the condor people visited the Vermilion Cliffs bordering Arizona's House Rock Valley just north of eastern Grand Canyon, they believed they had found the perfect spot for condors. I agreed with their assessment, but with one major caveat. My doubts about the success of this reintroduction effort, were based on the inadequacy of available food. I believed, perhaps incorrectly, that no areas remained where "natural" carrion could support a population of condors. The birds would have to be fed at feeding stations in perpetuity. If the "hand of human" was going to be forever the source of food for these giant buzzards, how could they be considered legitimate members of Colorado Plateau wildlife, or of any other ecosystem for that matter? Could any "wild" animal hold up its head in pride if it were forever dependent on the human animal for sustenance? I argued that perhaps the money and other resources

devoted to the condor recovery program could be spent on a cause that had a better chance of success. I once actually cited the condor as one of those endangered species we should abandon ("flush the species" was the unfortunate euphemism I think I used).

A single event changed my mind and softened my heart. I saw not just one but several California condors in the wild. Back in the early 1980s, before the last of the birds had been captured, I had made a special trip to the Sespe Wilderness, hoping to watch one of these birds soar into the sunrise, or sunset, or just to glimpse one of the creatures before they were gone. But it was not my time. Then in August of 1998, my son Tanner and I reached our favorite fishing camp seven miles up the Colorado River from Lees Ferry, to find the camp full—full of California condors. What a spectacle! These birds have a powerful effect on the human spirit. How truly magnificent they are. And to think that once I had written them off.

FOOD? SURELY THERE MUST BE ENOUGH FOOD. MAYBE THESE CONDORS WILL HAVE TO BE PROVISIONED FOREVER; IF SO, THEIR CONTINUED EXISTENCE IS WORTH IT.

Food? Surely there must be enough food. Maybe these condors will have to be provisioned forever; if so, their continued existence is worth it. But maybe they won't. Having met them face to face now many times over, and having watched for hours as they spread those gigantic wings (which can span more than nine feet) and soar among the cliffs, I am much more willing to believe that sufficient quantities of natural food are available to them. Greer Chesher cited Arizona Game and Fish statistics for deer, antelope, bison, and bighorn sheep densities in the vicinity of the Vermilion Cliffs release area, and, as best I can calculate, there should be at least three potential "prey" animals dying every day of the year, plus the occasional range animal. And when we talk of "vicinity" of the release area, we must remember that these birds can cover a lot of ground in a single day. Indeed, on August 23, 1998, three condors left the House Rock Valley Area and flew 250 miles to a spot in central Colorado where they rested for a couple of days while entertaining some very surprised tourists. They then turned around and flew back to the release site in less than twenty-four hours. Surely there must be enough food.

I FOR ONE, HUMBLY AND WITH APPROPRIATE CHAGRIN, AM CELEBRATING THE RETURN OF THE California condor to the Colorado Plateau, its ancient home.

Since the first release of six condors in December 1996 at Vermilion Cliffs, a total of twenty-eight 1.5- to three-year-old captive-bred condors have been returned to the wildlands of the Colorado Plateau. Of this number, only five have perished: three to coyotes, one to a golden eagle, and one to a collision with a power line. Remarkably, these birds are learning the social and physical skills (landing is a big deal for condors) needed to survive in the wild, and they are even starting to find food on their own. I for one, humbly and with appropriate chagrin, am celebrating the return of the California condor to the Colorado Plateau, its ancient home. To the dedicated members of the Peregrine Fund who are keeping track of these birds on a twenty-four-hour basis—thank you. (To biologist Mary Schwartz and her condor dog, I am the fisherman with the fisherson who gave you a boat ride out of the thunderstorm at Horseshoe Bend.)

PEREGRINE FALCON THE SUN HAD NOT YET PENETRATED THE DEPTHS of our Grand Canyon camp at river mile 56 near Kwagunt Rapids when my spotting scope picked out a tiny gliding form above the rim. It was May 1989 and I was part of an intensive study to determine the status of the peregrine falcon in the Grand Canyon. Falcon watchers, singly or in pairs, with a little food and a lot of high-resolution optics, were dropped off along the river at one-mile intervals on one day, then picked up at about the same time the next. Our task was simple: watch for peregrines and make detailed notes of where and when they were sighted and what they were doing. We were under contract with the National Park Service, and my co-workers up and down the river included some of the best-known names in raptor biology. "Cigar with golden wings," I shouted to my still sleeping colleague. We had a peregrine. When a peregrine falcon is glimpsed from over 3,000 feet below, all that can be seen, even with a powerful spotting scope, is a tiny cigar shape with two diminutive fins (wings). Backlit by the sun on that morning, the wings looked golden.

We watched the bird drift in and out of view, intermittently disappearing behind the rim some 3,200 feet above our heads. It appeared to be peacefully hunting. Peregrines, also known as duck hawks for their occasional preference for waterbirds, take all manner of bird prey in Grand Canyon, feeding quite commonly on white-throated swifts and violet-green swallows. These sparrow-sized birds would hardly seem a decent meal for such a high-level predator, but peregrines are frequently seen hunting in pairs, swooping into a flock of swifts and swallows, catching and eating several on the wing. On this particular morning, a flock of a hundred or more swallows was foraging at river level on millions of recently emerged midges. Suddenly the golden wings high above vanished and the cigar collapsed to a spot as the falcon plummeted straight toward the river. One-thousand-one. . . one-thousand-two. . . one-thousand-three. . . at the count of one-thousand-twelve, the peregrine, in an explosion of piercing wing shear, snatched a swallow fifty feet away from our position. Twelve seconds to travel 3,200 feet? Could this bird really have been traveling at upwards of 180 miles per hour? Maybe we counted wrong! I had heard the peregrine was fast, and that power dives in excess of 100 mph were credited to this sleek flying machine, but this was the first time I had personally witnessed such a display.

Before that study was over, we had documented over 100 pairs of peregrine falcons along the river and both rims of the park. This was astonishing for the time. The species was considered rare throughout the lower forty-eight states, such numbers were unprecedented in recent memory, and the number recorded in the canyon to that date had not exceeded four pairs. By 1989 the peregrine falcon was making a comeback, and the Grand Canyon was center stage.

Peregrine falcon. Photo by Marty Cordano.

A raven with smaller white-throated swift (left) and violet-green swallow. Illustration by Wood Ronsaville Harlin, Inc.

Peregrine falcon populations reached a low point in 1970 when only thirty-nine pairs were known to exist in the lower forty-eight states. All had disappeared east of the Mississippi River and 80 to 90 percent of previous populations had vanished from the West. In that year two subspecies of peregrines, the Arctic peregrine falcon and the American peregrine falcon, were listed as endangered under the Endangered Species Conservation Act of 1969, a precursor of the Endangered Species Act of 1973. Protection was extended when that more recent legislation was passed into law.

What caused the decline? It probably started in the 19th century. Creatures on the Colorado Plateau and elsewhere suffered from the prejudice of early settlers raising livestock and domestic fowl. Direct assaults were launched on all birds of prey, especially eagles and the larger hawks. Coyote poisoning also took its toll on unintended avian targets. To say that the birds of prey were unappreciated for their majesty and their role in balancing the ecosystem would be an understatement. Just about the time environmental awareness concepts were being unveiled by the early ecologists of the 1940s, World War II was upon us. Coincident with that war was the introduction of chlorinated hydrocarbon pesticides.

Today every junior high school student knows the story of DDT and its many disastrous effects. One of them erased birds from the sky. As the chemical accumulated up the food chain, it prevented normal calcium deposition during eggshell formation, rendering eggs too thin to sustain the weight of an adult bird attempting to incubate. Eagles and falcons were hit hardest, and while the chemical pervaded the environment, populations of these birds declined dramatically. Ecological science revealed the cause-and-effect relationship between DDT and disappearing raptors, but it took another decade before chlorinated hydrocarbon chemicals were banned unceremoniously in the United States and most of Europe.

Banning the poison, together with other recovery efforts, has restored the peregrine falcon and other birds of prey, notably the bald eagle, to the Colorado Plateau. Their increasing numbers here follow a national trend. So stunning is this recovery that in 1998 the U.S. Fish and Wildlife Service proposed removing the peregrine falcon from the endangered species list. Within the Rocky Mountain/Southwest recovery area, the goal had been to reach 183 pairs before considering delisting the bird. This area now has well over 500 pairs of breeding peregrine falcons. In this case, at least, equilibrium among species has been restored.

Bald eagle. ©Tom and Pat Leeson/DRK PHOTO.

AMERICAN BALD EAGLE

A MOUNTAIN LAKE—well, a reservoir actually—meanders through the forest not far from my home outside Flagstaff. Popular in summer with fishermen and boaters, upper Lake Mary in winter is a magnet for bird watchers. Almost every winter, like giant Christmas tree ornaments, migrating bald eagles decorate the pine trees along the shore, sometimes in clusters of thirty or more. Getting a good look at a lone bald eagle is reward enough for a day of birding (the fastest growing outdoor activity in America according to a recent survey), but when you can see several at a time, the experience is long remembered. The little knots of people and solitary watchers who line the lake, binoculars positioned in what appears to be permanent attachment to their heads, take these annual visitations for granted.

But not so long ago such eagle migrations were rare. And not so long ago I would have been stunned to see a bald eagle on my winter river trips through the Grand Canyon.

Bald eagles are back. Not just in northern Arizona, and not just on the greater Colorado Plateau, but across the nation. On July 12, 1995, a momentous day in the often bleak struggle to recover imperiled species, the bald eagle was downlisted from endangered to threatened, a big step on its journey of recovery from near extinction. Like the peregrine falcon and many other raptors, the bald eagle was hit hard by DDT. But this odious pesticide was only the straw that nearly broke the back of our country's avian icon. Some biologists have estimated that at the beginning of the 18th century the bald eagle population in the lower forty-eight states exceeded 20,000 pairs. By the early 1950s, shortly after the introduction of DDT, the eagle population had dropped to an estimated 10,000 nesting pairs. Deliberate shootings, dwindling food supply, loss of habitat, ingestion of pollutants, vehicular collisions, and electrocutions on power lines were primarily to blame. By the early 1960s, the effects of pesticide poisoning had kicked in, and a stunned biological community learned that no more than 450 nesting pairs remained. The bald eagle became one of the earliest entries on the nation's nascent endangered species list in 1967. Recovery has been slow and halting, but ultimately promising. Today, more than 5,300 nesting pairs embellish the skies south of Alaska.

Here on the Colorado Plateau we most often encounter bald eagles as winter migrants foraging in reservoirs like upper Lake Mary or in tailwaters below major dams. Just off the plateau to the south, along the Salt, Gila, Verde, and Bill Williams Rivers in Arizona, almost thirty pairs of bald eagles are known to nest, but within the plateau interior only a few breeding pairs are found. Bald eagles are primarily fisherbirds, though when fish are few they hunt small mammals and sometimes waterfowl. They will eat carrion, too. In fact, the great herds of bison that once blanketed America's plains probably supplied sufficient carrion to support huge numbers of bald eagles, condors, and other scavengers. On the Colorado Plateau, however, bald eagle numbers were probably never high. Before Euro-Americans began changing the landscape in a significant way, sources

Bald eagle in a snowstorm. ©Tom and Pat Leeson/DRK PHOTO.

Adult spotted owl with two owlets nesting in the cavity of a tree. Photo by Marty Cordano.

of fish were limited to the few streams with perennial flow and even fewer lakes. Carrion would have been plentiful, but nothing like that available on the Great Plains to the east. Ironically for the bald eagle as symbol of this nation's natural heritage, the landscape modifications that most severely disrupted ecosystems on the Colorado Plateau and throughout the Southwest—dams—have proven to be a boon to this species. Changing a stream or river into a lake has disastrous ecological consequences for most of the species dependent upon flowing water, but not every aspect of these changes is inimical to our native wildlife.

The bald eagle's apparent acceptance of habitat created by artificial lakes is, on the surface, something of a surprise. When we look closer, however, we find that these lakes are an abundant food source of mostly non-native fish—catfish, carp, bass, crappies—and that the clarity of standing water permits greater visibility and hunting success. In this instance, the bald eagle seems to have adapted fairly well to disturbances caused by humans.

There is a lot of talk about, and an ever-increasing level of support for, removing Glen Canyon Dam and returning the Colorado River to its original wild character. This is clearly a romantic notion, and may even have some long-term scientific merit. But one of the issues that will require resolution before such an action is seriously considered is its impact on endangered or threatened species like the bald eagle. There is no question that eventually dams like Hoover and Glen Canyon, and the rest of the many dams now blocking the free flow of the Colorado River through the heart of the plateau, will require decommissioning. In time they all will fill with sediment and no longer serve the purpose for which they were conceived. Nor will they support wildlife like the bald eagle. For now, though, as we watch the eagles congregate around these reservoirs, we should be pleased and relieved that these splendid birds have adapted so well to the situation, and that they, like us, have become functional members of these new, naturalized ecosystems.

MEXICAN SPOTTED OWL

OUR FAMILY HAS A SPECIAL RELATIONSHIP with the spotted owl. My youngest children, Cooper and Tanner, could produce a quality four-note spotted owl hoot before they could write a sentence. My two oldest, Carol Ann and Ken, from their late teen-age years ventured out of camp every summer night—setting out alone in the gathering dark to follow prescribed routes, hooting the owl song, in search of our special bird. We love this owl, and treasure those long, dark nights, but most of us can also dredge up memories of scalp-tingling fear when we share stories of our interactions with this species.

My family wasn't stalking the spotted owl for fun (although fun it often was). From 1989 to 1994, usually under contract with the U.S. Forest Service, I put together teams of biologists to actively search over 1.5 million acres of Colorado Plateau habitats for this stealthy denizen of the dark. My kids followed along, the older ones eventually becoming wildlife biologists in their own right.

To find owls, especially spotted owls, you must spend a lot of time in the deepest forests or remote and hidden canyons, and you have to be there in the dark. Moonless nights, when even shadows hide, are the best. It can get spooky out there in the wilderness with only a flashlight for a companion. My son Ken was bowled over by a black bear in the Gila National Forest early one morning

while "hooting" for owls. I suppose being mauled by a bear is bad at any time, but Ken assures me that the 2:00 A.M. darkness added to the terror of the moment. His wounds were superficial, and he still studies owls, but the incident gave him pause in his choice of careers.

My Navajo friend Bob Manygoats says owls, because they are creatures of the night and move with deathly silence, are considered by his people to be purveyors of bad luck and misfortune. A traditional Navajo doesn't even want to be around the feathers of these omens of darkness. The aversion most Native American groups share for owls, nocturnal owls especially, is easily understood once you've spent time searching the night for their presence. Many times I have been hidden in a copse of trees, using my vocal chords to mimic exactly the sound of a female spotted owl, only to look up and find an interested male sitting inches from my face. They hear your hoot and, instead of answering immediately, swoop in on wings that make no sound. With eyes and ears easily a hundred times sharper than our relatively weak human ones, owls own the night. We human animals, in contrast, seem to have a little section of DNA in our cells that, upon the approach of darkness, the real darkness of the forest, screams "Get out! Danger! Seek shelter!" In my opinion, owls have helped that little snippet of evolution along.

We know, of course, that spotted owls pose no threat to us. Quite the contrary. We monitor their numbers and distribution because we value the role they play in maintaining balanced forest ecosystems, and we are concerned about their survival as a species. Mexican spotted owls were officially listed as a threatened in April 1993. Reasons given included declining numbers and loss of habitat. Mexican spotted owls are known to use a variety of habitat types but are most often found in forests with high canopy closure, high stand density, a multi-layered canopy, uneven-aged stands, numerous snags, and downed woody matter. These conditions characterize old-growth mixed-conifer forests (usually more than 200 years old). Before 1991, management plans for most national forests occupied by the owl called for removal of old-growth timber in favor of even-age-stand silvicultural practices. Forest managers were maximizing the production of saw logs at the expense of ecosystem diversity.

Once the Mexican spotted owl was listed, every timber sale, every federal project having the potential to disturb spotted owl habitat, was required to demonstrate that owls would not be harmed. This resulted in many jobs for biologists and the protection of hundreds of pairs of spotted owls throughout the

Above: An immature spotted owl, still bearing its downy feathers. Below: Mexican spotted owl. Photos by Marty Cordano.

Bald eagle. © Tom and Pat Leeson/DRK PHOTO.

plateau. When an owl is found, over 600 acres of land are usually set aside as a protected core area, and another 2,000-plus acres around that core are restricted to only minimal disturbance. This amounts to a little over four square miles of protected habitat. Whenever an owl is located on the site of a proposed timber sale, that portion of the sale is canceled. This has made for some very tense confrontations between owl biologists and members of the logging community in rural areas throughout the plateau.

In the years I have spent communicating with this mottled grey-brown, dark-eyed owl (most owls have yellow eyes), I have learned some interesting lessons about the health of Colorado Plateau ecosystems. Some areas of the plateau are surprisingly blessed with spotted owls, while others, on the surface very similar, haven't supported a spotted owl in decades. In the Dixie National Forest of southern Utah, for example, we spent the summers of 1989-1992 scouring 250,000 acres of what should have been good owl habitat. In three years we heard only one spotted owl male, one time. Other contractors, working the steep and isolated canyons of the southern Utah rim country found an isolated owl or two, but our team worked an apparently suitable forest ecosystem and found nothing.

It wasn't until we had more experience, moving on to the Gila National Forest of New Mexico and the Uncompahgre National Forest of Colorado, that I finally understood what was going on. A night and early morning of walking in these dark forests was constantly punctuated by heart-stopping rustlings in the underbrush. There were also spotted owls, sometimes several. A night of walking the Dixie National Forest was like a night walking among the dead—it was tomblike in its silence. The difference was the relative abundance of nocturnal rodents. Spotted owls eat mostly small mammals when they can get them. If woodrats were sufficiently abundant, it is doubtful that any self-respecting spotted owl would ever eat anything else. When small mammals are scarce, the owls will even resort to tarantulas, scorpions and beetles for a meal, but this is fairly unusual. I have no real scientific data to back up my hypothesis, but my instincts tell me that livestock grazing pressure in the Dixie National Forest has been sufficient to remove the density of understory vegetation needed by the nocturnal rodents. Simply put, there wasn't, and still isn't, enough for the owls to eat in the Dixie.

THE OTHER ENDANGERED SPECIES of the Colorado Plateau (and there are many) are under threats that can be removed with some effort. Efforts are being made, but there is still much we need to know and need to do. Ongoing and future research on these and endangered species everywhere is critical to interpreting the warnings of Nature's messengers.

We remain very lucky, those of us who live on the Colorado Plateau, that the core of our greater ecosystem is still pretty much intact. I have found that watching and listening to the birds yields much, much more than pretty sights and sounds. So pick a bird, you will know the right one; let it be your spirit guide; listen to it, watch it, and wait. Gifts other than those ornithological will accrue. For me the California condor has become a spirit guide (I have others), reminding me of our human responsibility to take care of Mother Earth and her fragile inhabitants. Clearly, the condor's message to our species is "watch me, learn from me, help me, and just maybe—maybe there is sufficient time for you, by preventing needless extinctions, to learn enough to prevent your own."

STEVEN W. CAROTHERS began his professional career as a Grand Canyon field biologist for the Museum of Northern Arizona in the early 1970s. Today he is president and founder of SWCA, Inc., Environmental Consultants, and a trustee of the museum. His writings include the 1991 book *The Colorado River Through Grand Canyon: Natural History and Human Change* in which he and co-author Dr. B. T. Brown chronicled nearly three decades of ecosystem responses to Glen Canyon Dam. Dr. Carothers's current projects on endangered fishes in Grand Canyon follow his long-standing interest in finding the elusive balancing point between Glen Canyon Dam operations and the nurturing of native species in Colorado River habitats.

Egg hatching. Photo by Liz Hymans.

"…the Indian traders were the media through which were exchanged the values of two ways of life."

FRANK WATERS, MASKED GODS

The trading posts of the Four Corners area were outposts on the western frontier, centers of commerce and social interaction that bridged the worlds of the pioneer settlers and the indigenous peoples of the region. Their unique architecture reflects the landscape in which they were created: adobe or stucco, or simple wooden structures built of rough-hewn wooden beams, with dirt floors and lofty rafters from which many of the goods for sale were displayed. While they were centers of trade and social interaction, where the exchange of goods and services took place on a daily basis, they were also dwellings. They were home to the traders who traditionally formed close bonds with the land around them and its inhabitants.

Many of these outposts survive, in one form or another. And while the world of the traditional traders is all but gone, the vestiges that remain give us an idea of what that world must have been like. The more remote outposts seem to preserve a greater feeling of their original character. But even those close to today's commercial and tourist routes retain some lingering flavor of their past. Many continue to serve local populations, functioning as a primary source of food and supplies.

Cameron Trading Post, 1934 RB-MV 1201, Museum of Northern Arizona archives.

Tuba City Trading Post, 1932. Cline Library, Northern Arizona University, NAU.PH.413.696.

Catalogue and Price List

Navajo Blankets
& Indian Curios

J. L. HUBBELL
INDIAN TRADER
Ganado, Apache County, *Arizona*
Branch Store: Keam's Cañon, Arizona

Perhaps no single trading post captures the spirit of this past better than Hubbell Trading Post in Ganado, Arizona, preserved today as a national historic site. Managed by the National Park Service as a living and working trading post, to the casual observer little has changed since the Hubbell family turned it over to the park service in the 1960s. Behind the trading post itself may be found the dwelling of Lorenzo Hubbell, and the tiny hogan which served as a home for visitors and passers-by.

Much of this heritage is best preserved in historic photos, culled from photo archives throughout the region. But one photographer today, working in a medium as old as the buildings themselves, has captured the flavor, the essence, and the heritage of these historic dwellings on the Colorado Plateau.—GREER PRICE

Center, right: Original John Wetherill Trading Post at Kayenta, 1926, E-213W. Right: Harvey car at Tuba City Trading Post, 1927, E-213T. Museum of Northern Arizona archives.

Palladium Prints of Paula Jansen: The Trading Post Portfolio

Paula Jansen is a freelance photographer who specializes in palladium and food photography. "Platinum photography," she says, "is a 19th-century contact-printing process. Its rich tones resonate with an almost three-dimensional quality."

To create a palladium print, Paula starts by brushing an emulsion of metallic salts (either platinum or palladium) onto fiber-based paper. (Her brushmarks are still visible around the borders of the prints on the following pages.) Then she places a large-format negative, nearly the size of standard letter paper, atop the impregnated paper and exposes both negative and paper to ultraviolet light so that a positive version of the image is transferred to the paper. The exposed paper is immersed in a solution of organic salts until the image develops to the desired richness and range of tones, then it is bathed in diluted acids to stop the process. Each step of the process contributes to the character of the final image: amount of light exposure, exact composition of chemicals, timing, and, of course, the creation of the original negative. Paula captures her images for palladium printing with a large view camera, composing and exposing with the nature of the process in mind. The resulting archival prints are one-of-a-kind and are stable, permanent images that will not fade with age.

"After seeing Dick Arentz's platinum photographs in the 1980s," the photographer says, "I became smitten by palladium photography. I signed up for a workshop with Arentz and have been hooked ever since. Fifteen years later, mentored through Arentz's generous teaching spirit and inspirational art, I am still learning more and more about this beautiful process."

Hubbell Trading Post at Ganado, Lorenzo Hubbell seated in foreground. Museum of Northern Arizona archives.

The Gap Trading Post on the main road to Lees Ferry and Utah, 1926, E-213G. Museum of Northern Arizona archives.

2/28/99   1/50 Paula Jansen
Cameron Trading Post

Paula Jansen ¹/₅₀
Cameron Concho 2/881

Paula Jansen 1/50
Hubbell Rug Room

Cameron Trading Post

Paula Jansen 1/50
Cameron Grocery 2/28/90

Paula Johnsen 1/50
CAMERON DOOR 2/28/99

1/50 Paula Jansen

PERRY MESA LOOKING NORTHWEST FROM SQUAW CREEK.
BRADSHAW MOUNTAINS IN THE DISTANCE.

DAVID R. WILCOX,

GERALD ROBERTSON, JR.

AND J. SCOTT WOOD

In the structuring of human habitats, from single homesteads to group settlements, security and comfort have gone hand in hand. Consider the evolution of what we humans call home. By learning to control fire we invented the hearth, and the space around our hearths became home. The fire at the heart of our dwellings brought light, heat, and ease in cooking, but it also functioned as a weapon against feral marauders. Security remains integral to our homes and cultural landscapes. Its contemporary symbols are the locked door, car alarm, walled and gated communities, 911 rapid-response networks, and (on a global scale) nuclear deterrents. Whatever gives us a feeling of safety seems worth the price, even if it exaggerates social differences and creates boundaries that may increase the dangers we are so desperate to deny. The archaeology of this obsession with security poses interesting challenges.

Archaeology is fundamentally a study of the physical traces of past human behavior. These traces include artifacts, such as a superb yellowware pot from an ancient Hopi site or a beautifully crafted projectile point 11,000 years old. They also include standing architectural remnants and traces of surfaces such as floors, paths, or open courts—stages upon which the dramas of earlier lives were played out. These things are tangible. But equally important is the intangible: the record of where each artifact is found in relation to the sediments time has sifted over it, and where groups of artifacts are found in relation to each other and to the places where they were used. Ultimately, the primary data of archaeology is found less in objects than in relationships. The physical traces of past human behavior were all produced by the interactions of humans with one another and with their physical world. It is through the study of these relationships that we reconstruct the human past.

What relations can help us to understand the ancient need for security? What is the archaeological measure of fear? Asking this question in the American Southwest raises hackles, because it has long been thought that only peaceful relations prevailed here prehistorically. Unlike groups elsewhere in the world, the Pueblo and Piman Indians are uniquely held to be inherently peaceful; the word Hopi itself is said to mean peaceful. Most wish to believe that it is possible for humans to live together in peace and freedom, and to think that in the Southwest people once did so is greatly reassuring. But is it true, or is it a myth which denies our common humanity with prehistoric Southwestern peoples?

PERRY MESA, A 14TH-CENTURY GATED COMMUNITY IN CENTRAL ARIZONA

Very special thanks to Jerry Jacka for the aerial photographs he created to illustrate this story, and to Ed Campbell for piloting the aircraft from which he shot. Color maps were made by Chris Mitchell, starting from originals drafted by Jodi Griffith of the Museum of Northern Arizona.

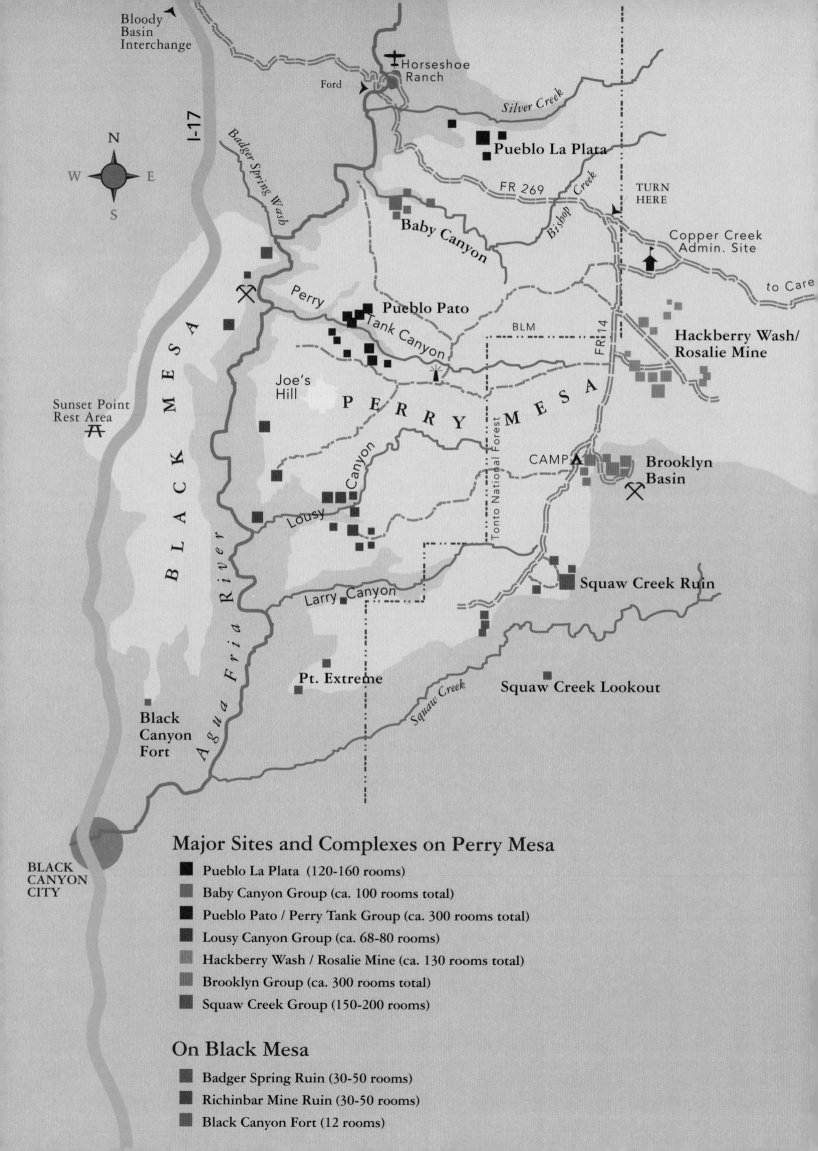

Bloody Basin Interchange

I-17

N
W — E
S

Horseshoe Ranch

Ford

Silver Creek

Pueblo La Plata

FR 269

Bishop Creek

TURN HERE

Copper Creek Admin. Site

to Care

Baby Canyon

Perry

Pueblo Pato

Tank Canyon

BLM

FR 14

Hackberry Wash/ Rosalie Mine

Joe's Hill

P E R R Y   M E S A

B L A C K   M E S A

Sunset Point Rest Area

CAMP

Brooklyn Basin

Lousy

Canyon

Agua Fria River

Tonto National Forest

Larry Canyon

Squaw Creek Ruin

Pt. Extreme

Squaw Creek Lookout

Squaw Creek

Black Canyon Fort

BLACK CANYON CITY

Major Sites and Complexes on Perry Mesa

- Pueblo La Plata (120-160 rooms)
- Baby Canyon Group (ca. 100 rooms total)
- Pueblo Pato / Perry Tank Group (ca. 300 rooms total)
- Lousy Canyon Group (ca. 68-80 rooms)
- Hackberry Wash / Rosalie Mine (ca. 130 rooms total)
- Brooklyn Group (ca. 300 rooms total)
- Squaw Creek Group (150-200 rooms)

On Black Mesa

- Badger Spring Ruin (30-50 rooms)
- Richinbar Mine Ruin (30-50 rooms)
- Black Canyon Fort (12 rooms)

In 1997 three archaeologists, of whom I am one, set out to answer a fundamental question regarding the nature of early relationships in the American Southwest. Our collaboration was born in a place called Perry Mesa, a grassy plateau fifty miles north of Phoenix. It is visible to the east across a thousand-foot-deep canyon from the Sunset Point rest area along Interstate 17. Distributed around the edges of Perry Mesa are the remains of a series of stone-masonry pueblos that, based on the pottery found there, date to the 1300s.

Scott Wood, archaeologist for the Tonto National Forest, has made detailed records of these sites over the past two decades. I am interested in his work, and in the area, because of my long study of regional warfare in the Southwest. The third man in our group, Jerry Robertson, alone among us has real military experience: he was a highly decorated captain in the 101st Airborne in Vietnam. One of his experiences was leading a "turned" group of Viet Cong behind the lines, where he learned first-hand what it takes to survive guerilla warfare. As Jerry and I worked together on various avocational-archaeology projects, the value of his military knowledge to our interpretation of archaeological sites became more and more clear.

ARCHITECTURE OF DEFENSE

When Scott, Jerry, and I began to pool our observations and ideas about the way its occupants inhabited this precipitously bounded landscape, almost immediately we noticed that the larger pueblo sites were patterned in their distribution. Perry Mesa is dissected by a series of side canyons. On every side canyon but one there were clusters of prehistoric dwellings placed as though to block access to the mesa from below. The largest, Squaw Creek Ruin on a wide bend of the watercourse it is named for, once had 150 rooms and was surrounded by a massive breast-high wall. Farther upstream a group of pueblos in Brooklyn Basin appeared to block access and, farther north along Hackberry Creek, there was yet another cluster of protectively placed sites. It looked as if each pueblo was positioned to protect the backs of its neighbors. Was Perry Mesa a militarily integrated system of settlements designed for defense?

Walking around the perimeters of individual sites, we noted other relationships that suggested defensibility. Baby Canyon Ruin is perched on an outcrop projecting over a canyon far below. Where the outcrop ends, a wall begins, as if to bar passage from below. Pueblo Pato has four structures, the

Putting the Past Behind Us

The question of prehistoric violence on the Colorado Plateau came to the attention of the public in November 1998 when *The New Yorker* magazine published a story by Douglas Preston on the work of Christy Turner, sensationally entitled "Cannibals of the Canyon." Turner's recent book (researched and written with his late wife Jacqueline) was entitled *Man Corn*, the Aztec term for human meat, and it set forth in excruciating detail the results of a 30-year study of mass interments whose characteristics appeared to Turner to provide evidence of cannibalism.

Pueblo people today are understandably distressed to hear such interpretations, and we may all sympathize with their dismay. Serious debate now centers on witchcraft as an alternative explanation for the specific features of the skeletal remains that seem to indicate cannibalism. In the scientific community, new data and more detailed studies will create a larger basis for consensus. In the public arena, scholars emphasize that evidence indicates only rare incidents of the acts under speculation, and that we have much to learn about their religious and ideological significance.

AT TOP: CHRISTY AND JACQUELINE TURNER IN 1957 AT THE MUSEUM OF NORTHERN ARIZONA.

BABY CANYON RUIN (ON THE KNOLL IN THE
FOREGROUND). PERRY MESA LOOKING WEST.
THE BRADSHAW MOUNTAINS ARE IN THE
BACKGROUND.

BABY CANYON RUIN (ON KNOLL IN CANYON).
PERRY MESA LOOKING EAST.

largest probably once two stories tall, arranged in a compact mass that would have been hard to attack. Positioned where vertical cliffs prevent its direct access, it provides an effective "window" from which to watch anyone coming up Perry Tank Canyon. Squaw Creek Ruin has an outer wall. Where the wall ends, crevices in the basalt outcrop have been filled in as if to prevent access from below. Anyone trying to enter from an unwalled access would be caught between the pueblo and the basalt outcrop at the end of the wall. Access from the east, south, and north is uphill, a decided advantage to the occupants. And if the wall were breached, those who stood behind it could withdraw to the pueblo, giving them what Jerry called "defense in depth."

DEFENSE AGAINST WHOM?

Having made these observations, we began to talk about the possible process of attack. Jerry pointed out that a defending commander would want advance warning of approaching marauders, so we began to look for a likely lookout post. Scott told us about a site he had found a few years earlier from a helicopter: a massive wall across a small, high point on the south side of Squaw Creek, within line of sight of the Squaw Creek Ruin. It was just the kind of site Jerry was proposing should be there!

Violence and Mayhem on the Colorado Plateau?

Evidence of prehistoric violence has long been known from archaeological sites on the Colorado Plateau. It includes projectile points embedded in human skeletal remains, heads buried without the remainder of the skeleton, and details of prehistoric architecture that seem to have been designed for defense. In their excavation of Cave 7 at Grand Gulch in the 1890s, the Wetherills uncovered a massacre of ninety Basketmaker people; in 1901, near Petrified Forest, Walter Hough found bones of prehistoric Puebloan peoples which seemed to indicate cannibalism. But archaeologists have only recently begun to form any kind of consensus that such evidence indicates widespread patterns of raiding or warfare.

The scholarly assembling of data, systematic analysis, and directed field studies have led to new proposals and increasing debate concerning the nature of prehistoric violence on the Colorado Plateau. One very positive result of all this has been the wide range of new questions we now ask about long-studied and familiar archaeological sites. Are the towers at Hovenweep part of a defensive line-of-sight system designed to guard and restrict access to precious canyon-head springs? Were the cliff dwellings at Mesa Verde arrayed so that their inhabitants might come to one another's defense? Were any of the "great houses" at Chaco fortresses?

Allowing for possibilities that once we would never have posed, we can now look at old dwellings with new eyes, in search of relationships we have not recognized before.

A few months later, in November 1997, we spent another three days together on Perry Mesa. On the way in, we stopped on the side of Bloody Basin Road to listen attentively to Jerry's ideas. "Why would anyone want to attack Perry Mesa? What did its people have that others would want?" There was no obvious answer to this question. Jerry pursued his thought. "Suppose then that the people on Perry Mesa were themselves the aggressors." He formed his hands into a circle. "Think of a castle," he said. "Now," moving his hands apart, "think of Perry Mesa as a castle." Suddenly, the vertical cliffs of the canyons became castle walls, and the pueblos the strong points around them.

We were proposing that Perry Mesa was, in modern argot, a "gated community." Gated against what? Against retaliation by the victims of attack? If the Perry Mesans were themselves the aggressors, the Phoenix Basin Hohokam would have been their most obvious target. This group had large villages, extensive irrigation agriculture, cotton-textile and sea-shell valuables, and may have been the wealthiest population in southern Arizona.

Picking up a handy stick from the side of the road, Scott and I drew a map in the dust, pooling our knowledge about the neighbors of the Perry Mesans. Who were friends, and who potential enemies? After extensive consideration, it seemed likely that the people living in Bloody Basin and along the upper part of the lower Verde Valley to the east were allied, and the Hohokam of the Phoenix Basin were the target of attack by Perry Mesans, if attackers they were.

LOOKOUTS AND CASTLE WALLS

Later that day, following up on Jerry's insistence that Squaw Creek Pueblo must have had some way to know of people approaching up Squaw Creek, we drove a nearly impassable road out to the southwest tip of Perry Mesa. No site was shown on the map Scott had given us, but, situated just as Jerry had predicted, we found a superb lookout-site: a twenty-room pueblo out on a point, with a perfect view of the place where Squaw Creek joined the Agua Fria River. Looking up the canyon, Scott could also see out on a point the lookout site he had found by helicopter. We later discovered that Museum of Northern Arizona staff had previously recorded the site, calling it Point Extreme. "Extreme" describes our feelings that day perfectly, as we came upon this confirmation of Jerry's hypothesis.

Over the campfire that night, we stayed up late talking about the behavioral processes of attack and defense. Given the early warning systems we had documented, and that were further indicated by extensive line-of-sight relationships among numerous of the pueblos documented by other researchers, Jerry pointed out that any attacker coming up from the south would have been strung out in the canyons, and thus all the more vulnerable to the Perry Mesan counter-defense. My earlier work had

suggested that the whole Phoenix Basin in the 1300s was politically integrated, binding together for common action a population of some 24,000 people. In theory, a retaliatory force from the Phoenix Basin to Perry Mesa could have numbered a thousand warriors. Could the mesa have defended itself against such a force? "Yes, indeed," Jerry said.

The next day we went over to the Hackberry Creek sites, and Jerry pointed out that the way they were deployed would have created a gauntlet, making it very difficult for an attacking force to get through. On a subsequent trip to Perry Mesa, we confirmed that the same is true of Brooklyn Basin. But Larry Canyon intrigued us. No pueblos were known to block access up this route, and at first this seemed a dangerous "back door" to have left open. Difficulties of access frustrated our own first efforts to investigate the problem. Scott then spotted a pueblo on an aerial photograph, and on a later trip we managed to get to this seven-room site at the top of a trail out of the canyon. There we found that early warning of approach via this route was available, but no large pueblos existed to block such access. We noticed, however, that forces from adjacent pueblos could readily have been deployed to repel an attack from that zone. The unique, large wall around Squaw Creek Pueblo may have been necessary because of the danger posed by this route. A force arriving on Perry Mesa at this point would have been easily surrounded from pueblos close at hand.

PETROGLYPHS AT SQUAW CREEK RUIN.

EARLIER DEFENSE SYSTEMS

Another question concerned the antecedents of the Perry Mesa system. Scott and I immediately thought about the string of hilltop sites in the foothills north of Phoenix, which extend from the Agua Fria near New River to the Verde River below Horseshoe Dam. They date to an earlier time and were generally thought to be "retreat" systems. In the flats below were a series of farmsteads or hamlets from which people retreated to the hilltops in times of danger. Clearly, this was a "system" organized very differently than was Perry Mesa. When we left Perry Mesa after our November trip, we wanted to know a lot more about its 14th-century neighbors and the earlier defense systems.

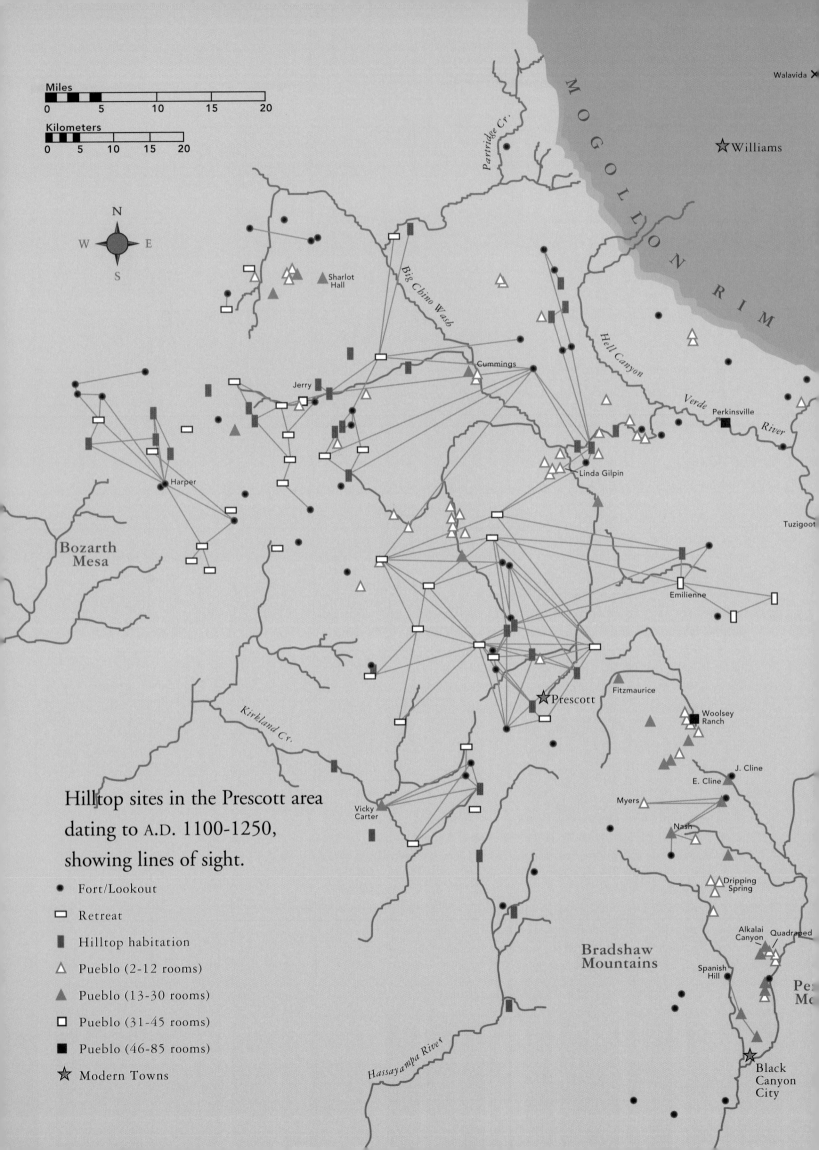

Hilltop sites in the Prescott area dating to A.D. 1100-1250, showing lines of sight.

● Fort/Lookout

▭ Retreat

▮ Hilltop habitation

△ Pueblo (2-12 rooms)

▲ Pueblo (13-30 rooms)

▱ Pueblo (31-45 rooms)

■ Pueblo (46-85 rooms)

☆ Modern Towns

Miles
0 5 10 15 20

Kilometers
0 5 10 15 20

N
W E
S

MOGOLLON RIM

Walavida

☆ Williams

Partridge Cr.

Big Chino Wash

Hell Canyon

Verde

Perkinsville

River

Cummings

Jerry

Sharlot Hall

Linda Gilpin

Tuzigoot

Harper

Bozarth Mesa

Emilienne

Kirkland Cr.

Prescott

Fitzmaurice

Woolsey Ranch

J. Cline

E. Cline

Myers

Nash

Vicky Carter

Dripping Spring

Bradshaw Mountains

Alkalai Canyon

Quadraped

Spanish Hill

Pe
Mo

Hassayampa River

Black Canyon City

EXTENDING THE SEARCH

Back at the Museum of Northern Arizona, I initiated a series of intense site-file searches and began to construct data tables and maps of all the 14th-century sites in west central Arizona, something that had not been done before. I also put together tables and maps of all hilltop sites that might be part of the earlier "retreat" systems. Field trips to three sites north of Sunset Point on the west edge of Black Mesa, which lies on the west side of the Agua Fria drainage opposite Perry Mesa, greatly enlarged our study area. Judy Taylor, an Arizona site steward who lives in Bumble Bee, told us about these sites,

LOOKING SOUTH ALONG THE CANYON OF THE AGUA FRIA RIVER.
PERRY MESA IS AT LEFT.

and we assumed they would date to the same time as the Perry Mesa pueblos. But this was not the case. They lacked the late pottery found at Perry Mesa and must have been part of an earlier system. Review of other known sites in the middle and upper Agua Fria drainage quickly revealed similar sites of comparable age. Suddenly the array of hilltop sites in the foothills north of Phoenix was joined continuously with the dozens of hilltop sites long known in the Prescott area, all apparently dating to the interval from A. D. 1100-1250. So we expanded our study area westward to the Big Sandy and northward to above the Mogollon Rim to consider all related phenomena in west central Arizona.

THIS HILLTOP RUIN, NEAR THE
SOUTH BANK OF NEW RIVER IN THE
AREA OF FIG SPRING, IS PROBABLY
THE TALLEST OF THE HILLTOP RUINS
NORTH OF PHOENIX, ARIZONA. THE
VIEW LOOKS WEST TO NORTHWEST.
THE BRADSHAW MOUNTAINS
APPEAR AT UPPER RIGHT.

HILLTOP PATTERNS

While many of Prescott's hilltop sites have been known since Territorial days, it was in the 1970s and 1980s that an avocational archaeologist named Ken Austin first systematically recorded them. Just as in the foothills north of Phoenix, small farmsteads or hamlets existed in the lower flats, and it has long been thought that in times of danger the people who lived there simply retreated to the hilltops for protection. But as we puzzled out the data, we began to question this assumption. First, not all hilltops were used in the same way: some seem to be lookouts, others have large compound walls and very few artifacts, while others have many artifacts and numerous rooms. Thus, while some of these hilltop sites were used for habitation, others might have had purely religious (or other) functions. Austin's studies of line-of-sight

LOOKING SOUTH ALONG THE CANYO
OF THE AGUA FRIA RIVER. PERRY
MESA IS AT LEFT.

THE SUN SPOTLIGHTS SQUAW CREEK RUIN (FOREGROUND) ON PERRY MESA.

among these hilltop sites leads us to propose that they can be grouped into a series of "local systems" that seem to form mutual protection associations: they could have provided early warning (or other information) to one another. Feuding among these local systems is one possible explanation for how such a settlement pattern could have evolved, over the century and a half it is thought to have existed.

The same conclusions can be reached for the hilltop sites north of Phoenix. Two other facts there, however, are of special note. First, the hilltop pattern first appears about A. D. 1100, just at the moment when irrigation communities near the foothills and throughout the whole lower Verde Valley were abandoned. Rich in saguaro, palo verde, and deer, the foothills became the domain of new people. The foothills were abandoned again about A. D. 1250, around the same time that large high walls or "compounds" were built in the communities on the Gila and Salt Rivers, placing household clusters behind protective walls. Buildings were also first erected atop "platform mounds," the public architecture of these communities. Clearly, major organizational changes were occurring in the Hohokam communities, and at the same moment that the people in the foothill zone moved away. We believe that these events are related.

Looking at the abandonment of the hilltop pattern in the Prescott area, which also occurred about A. D. 1250, we are led to an even more radical proposition. Suppose that small-scale raiding of Hohokam villages from the foothills north of Phoenix, the Agua Fria, and Prescott areas had been going on for many decades. Finally the Hohokam determine to stop it. If, like Kit Carson against the Navajo in 1863, they undertook a "scorched earth" policy, systematically burning fields and storage facilities at harvest time, moving fast and then withdrawing, they could have brought about a region-wide abandonment due to famine. We know of no other process that can explain the apparent contemporaneity of this abandonment of so large an area. We suspect they then moved eastward into the Verde Valley and northward onto Perry Mesa.

On a return trip to Perry Mesa we found that, at many of the pueblo sites, there were earlier components that date to the late 1200s. The defensive system represented by the pueblos was thus apparently in place by that time, and was further "hardened" by the construction of the pueblos a little later. We suggest that what had been a low level of conflict aggravated by raiding escalated in the late 1200s into warfare between much larger groups, involving hundreds of warriors. As the aggressors, the Perry Mesans may have begun to raid not only for small amounts of resources like corn or cotton textiles, but for captives, and to kill anyone who got in their way. Retaliation, too, would have been much fiercer, and deflecting it would have required greater coordination and social planning. Military integration, and probably political integration, would have been the result. Squaw Creek Pueblo, which is behind the lines, as it were, was probably the seat of the "command structure" for this "polity," the place that could receive information and send out commands.

Are the towers at Hovenweep part of a defensive line-of-sight system designed to guard and restrict access to precious canyon-head springs? Were the cliff dwellings at Mesa Verde arrayed so that their inhabitants might come to one another's defense? Were any of the "great houses" at Chaco fortresses?

With the assistance of Peter Pilles, archaeologist with the Coconino National Forest who also has a long-standing interest in the archaeology of Perry Mesa, we succeeded in assembling a comprehensive map of all known 14th-century sites in west central Arizona. We then solicited help from many other people in constructing a comparable map for east central Arizona as well. We were looking for a pattern of settlement clusters separated by areas without sites that formed "no man's lands," in order to test our ideas about regional warfare. Such a region-wide pattern is known during the early Protohistoric period (A. D. 1450-1600) and is now well established above the Mogollon Rim on the southern Colorado Plateau during the 14th and early fifteenth centuries. By assembling settlement data for sites below the rim, we wanted to see if similar patterns held there, as well.

In east central Arizona the same pattern does hold; along the Verde River, from Perkinsville to Davenport Wash, and from Perry Mesa to Polles Mesa (on the East Verde), it does not. As we had supposed that day out on Bloody Basin Road, Perry Mesa appears to have been closely allied with its neighbors to the east and northeast, a confederation we call the "Verde Alliance." From room counts we estimate it included twelve thousand people, three thousand of them on Perry Mesa. Between them and the Phoenix Basin Hohokam to the south is a wide no-man's land, and there is another such no-man's land, nearly as wide, between it and Chavez Pass to the northeast. Between Polles Pueblo and Rye Creek Pueblo in the upper Tonto Basin there were once villages and lookouts on high pinnacles. But they were abandoned by A. D. 1300, and a wide no-man's land opened up between these sites. East of Reno Pass another opened up between the lower Tonto Basin and the Phoenix Basin (or the lower Verde).

A PICTURE OF 14TH-CENTURY ARIZONA

Starting on Perry Mesa with the idea that the people there had deployed themselves to protect each other's backs and examining information collected over the past 100 years, we were to construct a general picture of 14th-century central Arizona. It is a picture of multi-settlement clusters bounded by no-man's lands which demarcate the clusters as polities at war with one another. Population estimates for these clusters range from 500 to 3,500 people; data currently available suggest that some shrank while others grew larger. By A. D. 1400-1450 all were abandoned. We propose that regional warfare was a principal reason for this.

Some archaeologists will disagree with this conclusion. We are not troubled by that, for our view of science is that it proceeds as a series of controversies among contending positions, controversies that facts eventually (if only partially) resolve. That happens as independent observers reach common conclusions about certain points. Facts are not easy to come by, and we are only at the beginning of the study of southwestern prehistory. As more relationships are documented, all of us may have to change our conclusions. But we hope that others will agree that the collaboration between archaeologists and those with real military experience has

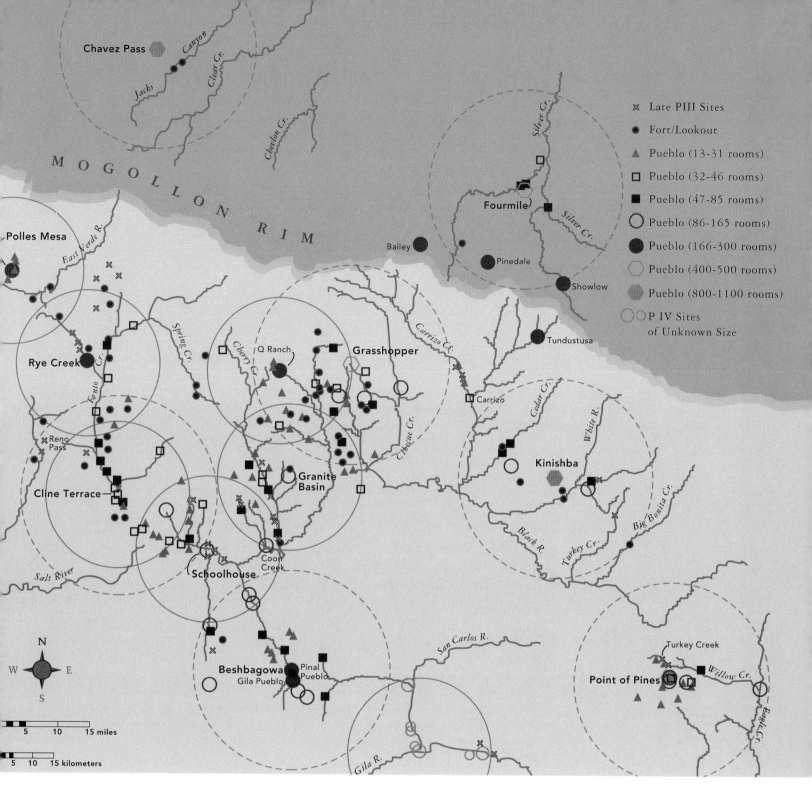

14th-Century Pueblos in East-Central Arizona

CIRCLES INDICATE AREAS AROUND MAJOR SITES TRAVERSABLE IN ROUGHLY A DAY; LARGER
CIRCLES DENOTE 15-MILE DISTANCES FROM THE CENTRAL POINT AND SMALLER ONES 11-MILE
DISTANCES. THE CIRCLES DEFINE SITE CLUSTERS AND SUGGEST POLITICAL BOUNDARIES. THE
LOCATIONS OF LATE PUEBLO III SITES THAT WERE ABANDONED BY AROUND A.D. 1300 AND THE
POSITIONING OF FORT/LOOKOUT SITES REINFORCE THE INDICATION OF BOUNDARY ZONES WHICH
BECAME "NO MAN'S LANDS."

been fruitful. The military mind thinks in terms of relationships, and the study of relationships, we believe, are key to understanding the past.

While we still may be a long way from achieving peace on earth and justice for all, we suggest that thinking dynamically in terms of interactions, or relationships, holds out the best hope for a level of understanding that can lead to peace. Let us not be blinded by insisting on static categories imposed by one side or another. Let people be people, and let us see them as they are, not as we wish them—or fear them—to be. We then can move forward with a wider view, basing our choices for the future on a clear understanding of our past.

Authors

David R. Wilcox is senior research archaeologist and special assistant to the senior vice president at the Museum of Northern Arizona. For nearly thirty years he has worked in Southwestern archaeology. He has published eleven books and sixty-five articles on subjects as diverse as Hohokam ballcourts, the political organization of the Chacoan system, and the history of Southwestern archaeology. He especially enjoys working with avocational archaeologists who bring fresh perspectives and sensible judgement to the study of the archaeological record.

Gerald Robertson, Jr. served in Vietnam as commander of an infantry rifle company, 101st Airborne, with the rank of captain. Before his discharge in 1968 he was awarded a Silver Star, three Bronze Stars, an Air Medal, and two Purple Hearts. Since 1992 he has worked with Dr. Wilcox, first as a volunteer and later as president of the Verde Valley chapter of the Arizona Archaeological Society. He lives in Sedona.

J. Scott Wood is forest archaeologist for the Tonto National Forest, which includes the eastern portion of Perry Mesa. In his ongoing attempt to understand the adaptation of humans to the central Arizona landscape, he has focused on the development of the Hohokam, Salado, Sierra Anchan, and Prescott traditions, the study of prehistoric ceramics, the prehistory and history of the Western Apache and Yavapai people, historic Anglo settlement history, and the military relationship between the U.S. Army and the Tonto Apache. His interest in warfare actually derives from a long-term interest in the history and archeology of Roman and Anglo-Saxon England. His association with avocational archeologists in general and the Arizona Archeological Society in particular goes back many years and involves a number of survey and excavation projects. Working with the Bureau of Land Management, he hopes one day to develop many of the Perry Mesa sites for public interpretation.

SUGGESTIONS FOR FURTHER READING

"The Scream of the Butterfly" by David R. Wilcox and Jonathan Haas, in *Themes in Southwest Prehistory,* edited by George J. Gumerman, 1994, School of American Research Press, Santa Fe.

Man Corn: Cannibalism and Violence in the Prehistoric American Southwest by Christy G. Turner II and Jacqueline A. Turner, 1998, University of Utah Press, Salt Lake City.

Prehistoric Warfare in the American Southwest by Steven A. LeBlanc, 1999, University of Utah Press, Salt Lake City.

LOOKING SOUTH TOWARD PERRY MESA (DISTANT LEFT). NEW RIVER MOUNTAINS AT FAR LEFT.

JOINT PUBLISHERS

MUSEUM OF NORTHERN ARIZONA
3101 N. Ft. Valley Rd., Flagstaff AZ 86001
Tel. 520-774-5213

This private, nonprofit institution is dedicated to the understanding, interpretation, and preservation of the Colorado Plateau. It offers the ideal introduction to the fine arts, Native American cultures, and natural sciences of the region through exhibits, programs, and outdoor excursions. Over its 70-year history, MNA has evolved into an intergenerational campus that includes a high school, residential community, and senior living center.

GRAND CANYON ASSOCIATION
P.O. Box 399, Grand Canyon, AZ 86023
Tel. 520-638-2481
www.grandcanyon.org

Grand Canyon Association, a non-profit organization established in 1932, exists to support education, research, and other programs for the benefit of Grand Canyon National Park and its visitors. GCA publishes books and other materials relating to Grand Canyon and the surrounding region. Funds generated by the association are used to support the educational goals of Grand Canyon National Park.

PLATEAU PARTNERS

ARCHES NATIONAL PARK
P.O. Box 907, Moab, UT 84532
Tel. 435-259-8161 (voice)
Tel. 435-259-5279 (TTY)

Arches National Park contains the world's largest concentration of natural stone openings. Red sandstone cliffs, spires, windows, and walls delight the eye and challenge the imagination. The park offers a great family experience. Paved roads guide visitors to spectacular views, and short easy hikes lead to fanciful formations.

ARIZONA STRIP INTERPRETIVE ASSOCIATION
345 East Riverside Drive, St. George, UT 84790   Tel. 435-628-4491

The Arizona Strip Interpretive Association (ASIA) supports the Arizona Strip, a vast land stretching from the Nevada border to the Colorado River and from the Utah border to the Grand Canyon, boasting 4,000 miles of unpaved roads. Sales merchandise and Brown Bag Lectures are featured at the Bureau of Land Management (BLM) Interagency Information Center in St. George, Utah. The center is the place to buy maps, posters, and books concerning the Arizona Strip, Pine Valley Forest, Dixie BLM, Colorado Plateau, state and federal parks, and all recreational activities. Annual membership is $10.

BRYCE CANYON NATURAL HISTORY ASSOCIATION
Bryce Canyon National Park
Bryce Canyon, UT 84717
Tel. 435-834-5322

The mission of Bryce Canyon Natural History Association is to assist and promote the historical, scientific, and educational activities of Bryce Canyon National Park. It also supports the research, interpretation, and conservation programs of the National Park Service.

CANYONLANDS NATURAL HISTORY ASSOCIATION
3031 South Highway 191, Moab, UT 84532
Tel. 435-259-6003

Canyonlands Natural History Association is a not-for-profit cooperating association operating in southeast Utah. CNHA is a partner to the land management agencies there in the development and marketing of high-quality interpretive materials which help to educate the visiting public as to sustainable land ethics.

CAPITOL REEF NATIONAL PARK
HC 70, Box 15, Torrey, UT 84775-9602
Tel. 435-425-3791

Capitol Reef National Park encompasses the Waterpocket Fold, a huge monocline eroded into a spectacular jumble of cliffs, domes, spires, canyons, and arches. The Fremont River cuts through the fold, supporting a diversity of plants and wildlife. Fremont Indian petroglyphs, historic structures and fruit orchards planted by Mormon settlers are preserved in the Fruita Historic District.

CAPITOL REEF NATURAL HISTORY ASSOCIATION
HC 70, Box 15, Torrey, UT 84775
Tel. 435-425-3791

Capitol Reef Natural History Association promotes the historical, cultural, scientific, interpretive, and educational activities and research of Capitol Reef National Park through the donation of proceeds from the sale of interpretive materials. Sales items are available at the Visitor Center and the newly renovated and refurnished Gifford farmhouse. The farmhouse, where visitors can purchase replicas of the past made by local artisans, serves as a cultural demonstration site to interpret the early Mormon settlement of the Fruita Valley.

COLORADO NATIONAL MONUMENT ASSOCIATION
c/o Colorado National Monument, Fruita, CO 81521   Tel. 970-858-3617

Since 1964, Colorado National Monument Association (CNMA) has enhanced the visitor experience at Colorado National Monument, providing a marketplace for educational materials and vital financial support for research, interpretation, outreach, and staffing. CNMA publishes information about the northeast edge of the Colorado Plateau. CNMA members enjoy benefits that include a bimonthly newsletter, discounts, and invitations to special events.

DINOSAUR NATURE ASSOCIATION
1291 East Highway 40, Vernal, UT 84078-2830   Tel. 435-789-8807

The Dinosaur Nature Association (DNA) is a not-for-profit organization created to aid the interpretive, educational, and scientific activities of the National Park Service at Dinosaur and Fossil Butte National Monuments. For more information or a catalog call 1-800-845-DINO (3466) or write to the above address.

DIXIE INTERPRETIVE ASSOCIATION
1696 Tamarisk Drive, Santa Clara, UT 84765
Tel. 435-628-3969

The Dixie Interpretive Association provides interpretive materials to increase awareness about multiple use management on the Dixie National Forest. Eighty percent of revenues go toward interpretive/educational programs and projects. Memberships are available. Association sales outlets are located in district offices in Cedar City, Panguitch, Escalante, and Teasdale, Utah. Visitor centers are located at Red Canyon (on Highway 12), Duck Creek (on Highway 14), and Wildcat (on Highway 12 south of Torrey).

ENTRADA INSTITUTE / FRIENDS OF CAPITOL REEF
P.O. Box 750217, Torrey, UT 84775
Tel. 435-425-3265

The Entrada Institute / Friends of Capitol Reef celebrates the human and natural history of the Capitol Reef environs through the arts, humanities, and sciences. The institute sponsors a variety of courses for students of all ages and encourages a program of lifelong learning for seniors. Members receive discounts on classes, mailing, and a subscription to a semiannual newsletter.

GLEN CANYON NATURAL HISTORY ASSOCIATION
P.O. Box 581, Page, AZ 86040
Tel. 520-645-3532

We are dedicated people working in partnership to promote stewardship and inspire public awareness of resource issues affecting the Colorado Plateau. We join together for the support of educational, historical, and research projects within Glen Canyon and surrounding federal lands.

GRAND CANYON NATIONAL PARK
P.O. Box 129, Grand Canyon, AZ 86023
Tel. 520-638-7888

The Grand Canyon, a World Heritage Site, is recognized as a place of universal value, containing superlative natural and cultural features that are being preserved as part of the heritage of all the people of the world.

GRAND CANYON TRUST
2601 N. Fort Valley Road, Flagstaff, AZ 86001
Tel. 888-GCT-5550
www.grandcanyontrust.org

The mission of the Grand Canyon Trust is to protect and restore the canyon country of the Colorado Plateau—its spectacular landscapes, flowing rivers, clean air, diversity of plants and animals, and areas of beauty and solitude. For more information on the Trust and our work, please call our toll-free number, visit our website, or write to us as the address above.

HUBBELL TRADING POST NATIONAL HISTORIC SITE
P.O. Box 150, Ganado, AZ 86505
Tel. 520-755-3475

Hubbell Trading Post National Historic Site is the oldest continuously operating trading post in the Navajo Nation and is the best remaining example of a traditional Southwest trading post. A crossroads of culture, the site preserves the past through its historic structures, Native American arts and objects, the landscape, and customs of another time.

AB NATIONAL FOREST
800 6th Street, Williams, AZ 86046
Tel. 520-635-8200

The Kaibab National Forest is an integral part of the heart of the Colorado Plateau. It surrounds the Grand Canyon and the communities of Williams, Parks, Tusayan, and Fredonia. One current focus is building relationships with the people interested in stewardship of these lands.

MESA VERDE MUSEUM ASSOCIATION
P.O. Box 38, Mesa Verde , CO 81330
Tel. 970-529-4445

The Mesa Verde Museum Association, a not-for-profit organization authorized by Congress and established in 1930, assists and supports the various interpretive programs, research activities, and visitor services of Mesa Verde National Park and Hovenweep National Monument. MVMA offers a membership program and supports the National Parks Electronic Bookstore at www.npeb.org

MESA VERDE NATIONAL PARK
P.O. Box 8, Mesa Verde National Park, CO
81330   Tel. 970-529-4475

Mesa Verde National Park, established in 1906, is one of the world's premier archaeological areas because of its high concentration of mesa-top structures and spectacular cliff dwellings. The park preserves the cultural heritage of the Ancestral Puebloan people, who lived in the area from A.D. 550 to A.D. 1300.

NORTHERN ARIZONA UNIVERSITY, CLINE LIBRARY
Box 6022, Flagstaff, AZ 86011-6022
Tel. 520-523-5551
www.nau.edu/library/speccoll/

As part of an educational institution, the Cline Library serves the Northern Arizona University academic community and the public by providing support for curricular, information, and research needs. The library's Special Collections and Archives Department collects, preserves, and makes available archival material which documents the history and development of the Colorado Plateau from prehistory to the present. The rich holdings represent a variety of disciplines and formats.

PEAKS, PLATEAUS AND CANYONS ASSOCIATION
c/o Tracey Hobson at Mesa Verde Museum Association, P.O. Box 38
Mesa Verde, CO 81330 Tel. 970-529-4445

Linked by both terrain and mission, the Peaks, Plateaus and Canyons Association (PPCA) is a group of not-for-profit educational associations, museums, and federal/state land management/tourism agencies. PPCA members seek to heighten visitor appreciation and understanding of the varied resources of the area on and around the Colorado Plateau. PPCA exchanges information and underwrites projects that advance the collective mission.

PETRIFIED FOREST NATIONAL PARK
P.O. Box 2217, Petrified Forest, AZ 86028-2217
Tel. 520-524-6228

Petrified Forest National Park is a respite in time where fossils of Triassic creatures and petrified wood converge with archaeological sites and historic settlements. This arid plateau also reveals the diversity of color and texture which are trademarks of the Painted Desert landscape. The park is a fascinating place where an ancient past meets the modern world.

PETRIFIED FOREST MUSEUM ASSOCIATION
P.O. Box 2217, Petrified Forest, AZ 86028-2217
Tel. 520-524-6228

Petrified Forest Museum Association is a not-for-profit organization established in 1914 to assist the interpretive, resource management, and educational programs at Petrified Forest National Park. Proceeds from the sale of publications are used to provide free information handouts, support scientific research, environmental education, Jr. Ranger programs, and special events.

USDI BUREAU OF LAND MANAGEMENT
Utah State Office, P.O. Box 45155, Salt Lake City, UT 84145   Tel. 801-539-4223

The United States Department of the Interior's Utah Bureau of Land Management (BLM) is a federal land management agency responsible for managing, protecting, and improving 22 million acres of public lands in the state. Resources contained on these public lands include recreation, range, timber, minerals, watershed, fish and wildlife, wilderness, air, scenic, scientific, and cultural values. BLM is committed to caring for these resources in such a manner as to serve the needs of the American people for all times.

ZION NATURAL HISTORY ASSOCIATION
Zion National Park
Springdale, UT 84767   Tel. 435-772-3264

Zion Natural History Association (ZNHA) is a not-for-profit corporation working in cooperation with the National Park Service. The association funds interpretive projects, scientific research, and free publications for park visitors through sales of publications, maps, and other interpretive items and members' support. ZNHA members receive a discount on sales items and a semi-annual newsletter.

Packrat nest enthroned in an abandoned outdoor privy. Photos by Paul Berkowitz.